DATE DUE

OC 1 1 '95			
MY 1 9 '96			
1 0 '97			
JY 1 2 '97			
OC 07 '99			
10/19/92			

DEMCO 38-297

SCIENCE TOYS

Laurence B. White, Jr.

drawings by

Marc Tolon Brown

Addison-Wesley

Books by Laurence B. White, Jr.

Science Games
Science Puzzles
Science Toys
Science Tricks
Investigating Science with Coins
Investigating Science with Nails
Investigating Science with Paper
Investigating Science with Rubber Bands
So You Want to be a Magician?
The Great Mysto . . . That's You

An Addisonian Press Book

To Dave . . .
who guarantees every experiment
because they worked for him

Addison-Wesley Publishing Company, Inc.
Reading, Massachusetts 01867
Printed in the United States of America
First Printing

WZ/WZ 08598 12/75

Library of Congress Cataloging in Publication Data

White, Laurence B
Science toys.

SUMMARY: Directions for some simple projects, such as a mold garden, a shadow clock, and an air spinner, that demonstrate basic scientific principles.
1. Science—Experiments—Juvenile literature.
[1. Science—Experiments] I. Brown, Marc Tolon, illus. II. Title.
Q163. W489 507'.2 74-2138
ISBN 0-201-08598-4

Toys are to Make

Toys are made to play with.
So are these things toys?

- a paper flower that blooms in a glass of water.
- A ghost that sticks to the wall when you rub it.
- A balloon dog that rolls on his back at your command.

They are all toys, but they are a special kind.
Science makes them work.
It makes the flower bloom, the ghost stick, and the dog roll.
You'll find here lots of toys like these.
You need just a little science to start toys working for you.

A Flower That Blooms

Cut a flower out of a paper towel.
Make it about this big.
Fold the petals down.

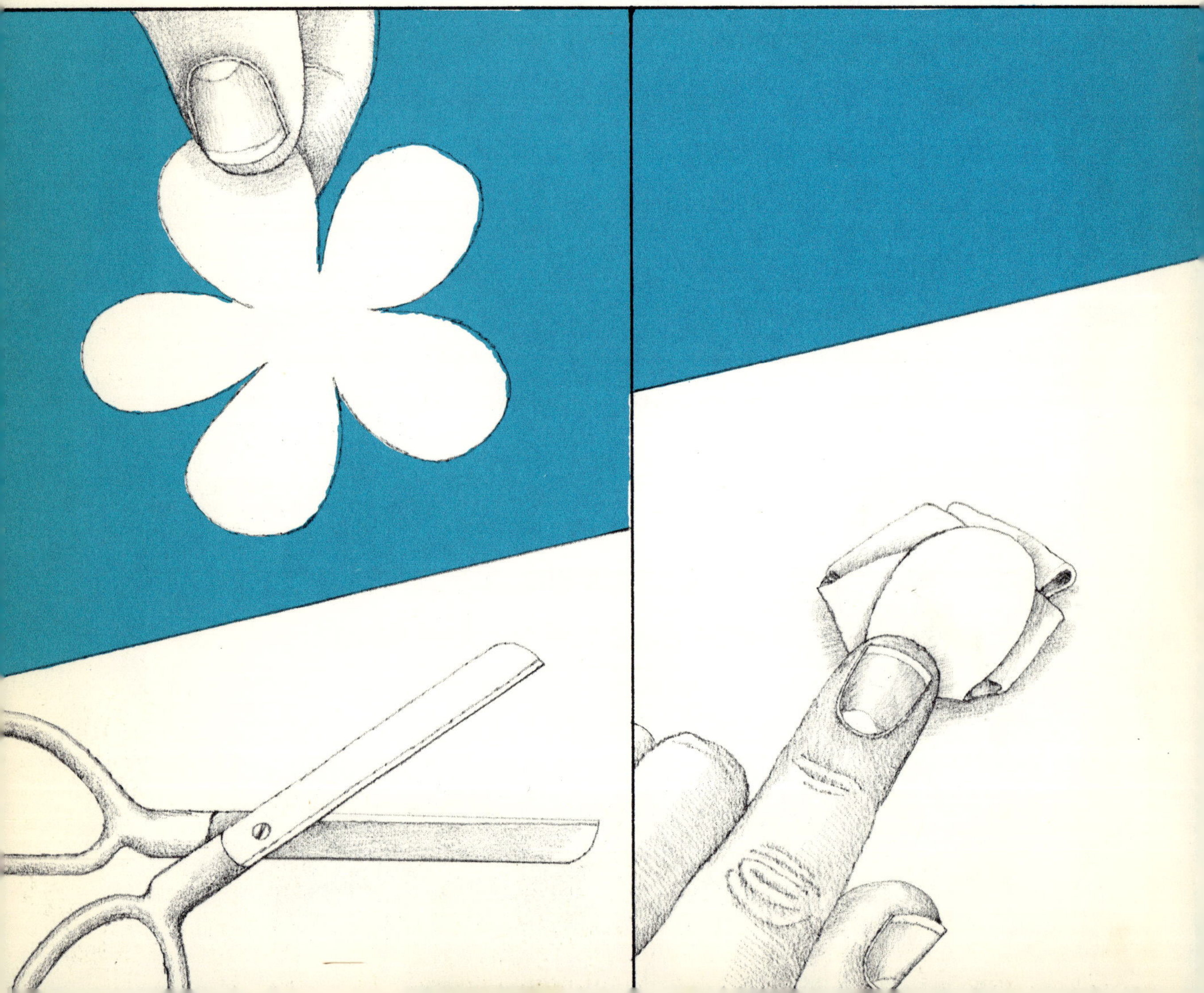

Drop the folded flower into a glass of water.
The paper soaks up some water.
The water makes the paper swell up.
Quickly . . . the petals open.
And the flower blooms.

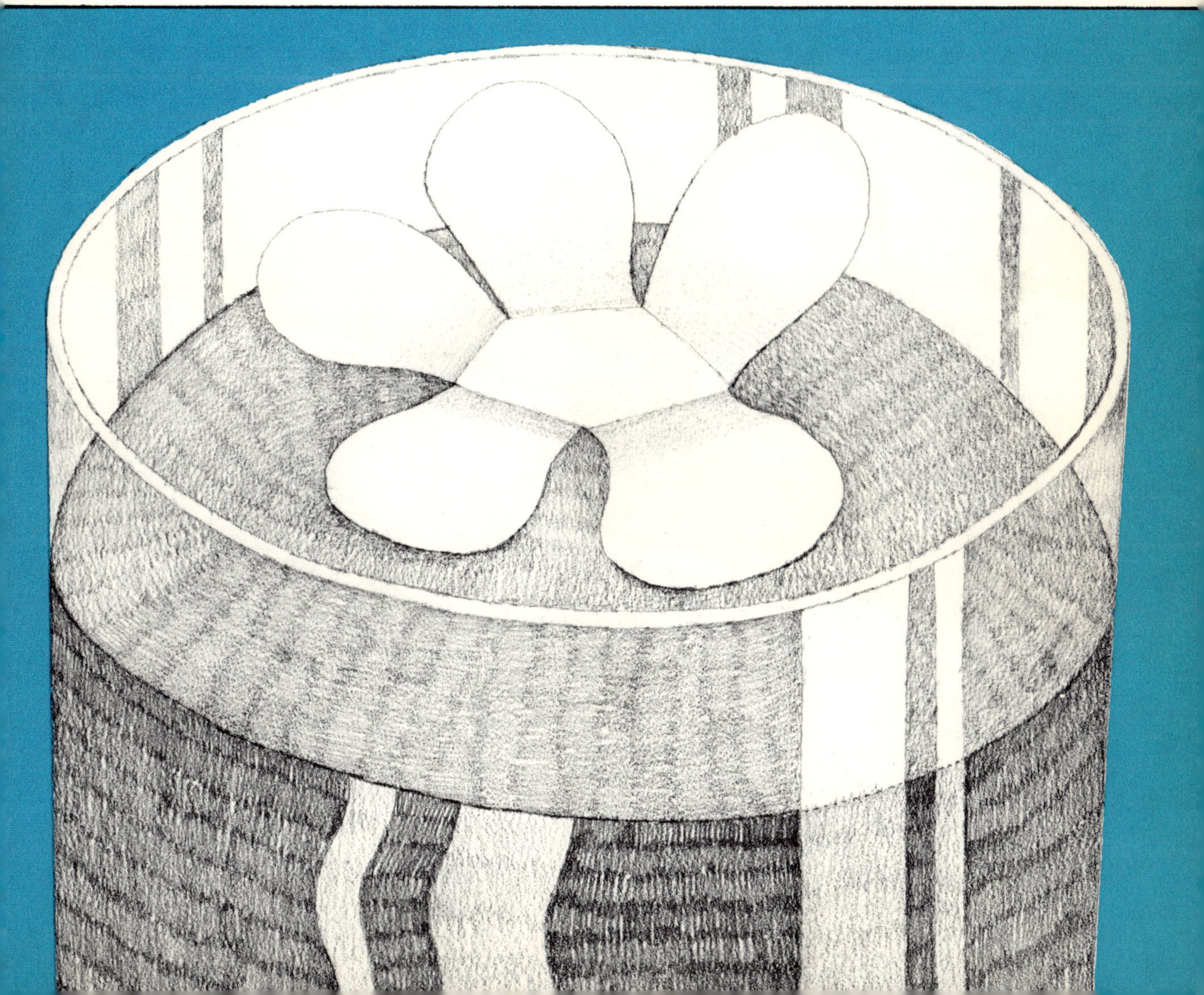

A Magic Card

Take a piece of cardboard.
Draw some straight lines across it.
Make the lines dark.
Push a pencil through the middle.
Spin the card on the pencil.
Look!
The straight lines change to curved lines.
The card is not really magic.
Straight lines just look curved,
because they are going around.

A Man That Moves

Fold a long strip of paper in half.
Draw a man on the bottom.
Draw another man on top.
Make their arms and legs different.
Hold the top sheet by one corner.
Flick it up and down quickly.
You will see the man move.
He seems to move his arms and legs.

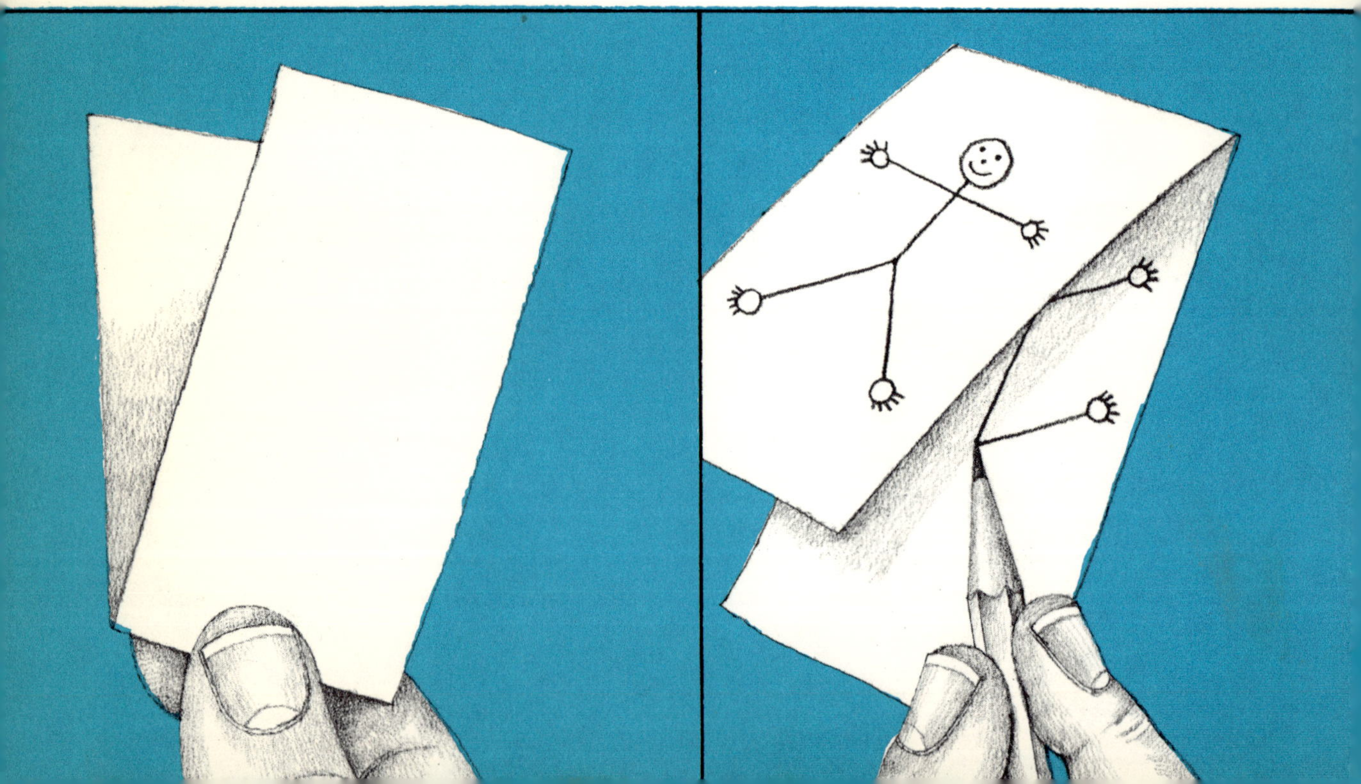

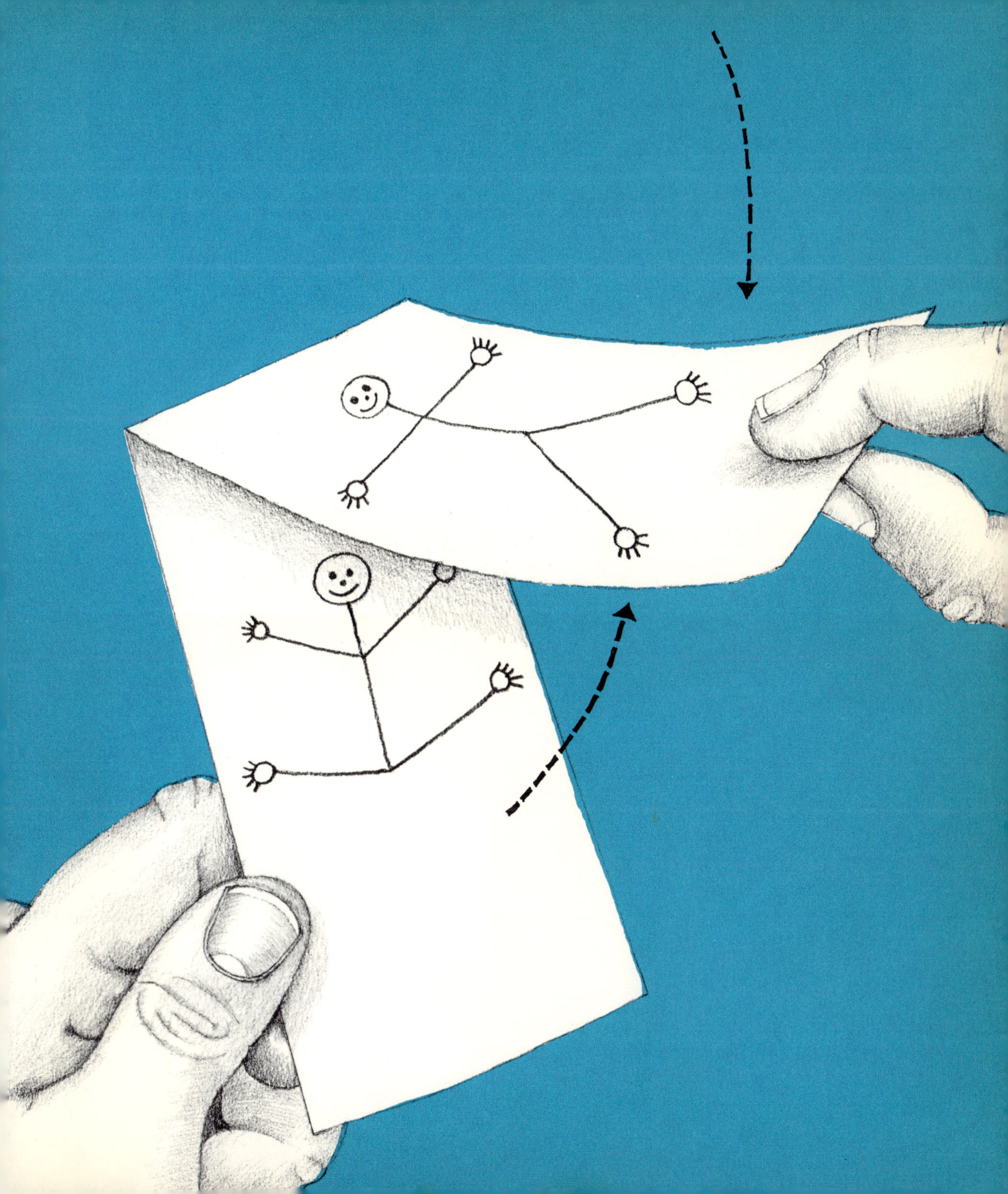

Fold a Fan

Fold a big piece of paper.
Fold it back and forth,
and back and forth again and again.
Hold it by one end.
Open the other end out.
Fan your face.
You cannot see air.
Or touch it or taste it.
But you feel it when it moves.
On a hot day fanning your face certainly feels good.

Fold a Spooky Screamer

Cut a piece of paper about this big.
Fold it in three places.

Hold it between your fingers.
Put your lips right against it.
Blow hard in the middle.
Blowing air can make spooky sounds.
Your breath is a little wind.
It will make the paper scream!

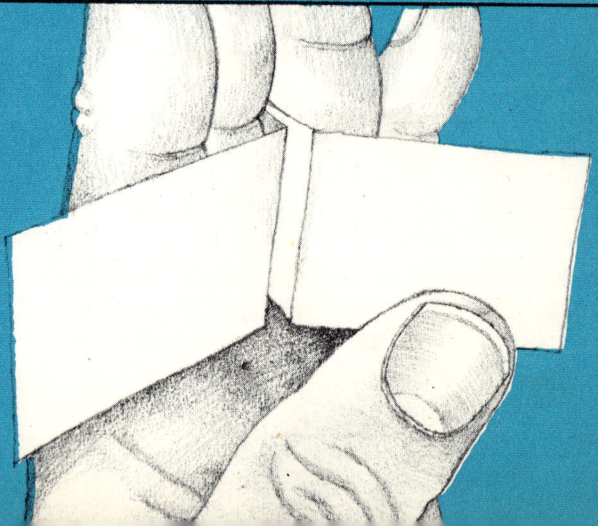

EEEEEKKK!!!

Fly an Air Spinner

Cut a long strip of paper.
Fold the two flaps down.
One one way, one the other way.
Put a paper clip on the bottom.
Drop the folded paper.
The air pushes on it.
Air makes it move.
It spins as it floats down.

A Strange Airplane

Cut two strips of heavy paper.
Tape them in loops.
Tape the loops to a drinking straw.
Toss it in the air.
It flies!
It has no wings or tail,
but the air holds it up.
Airplanes can be different shapes.

A Water Magnifying Glass

Cut a big hole in a piece of cardboard.
Cover it with clear plastic food wrap.
Lay it on a picture in a newspaper.
Let one drop of water fall on the plastic.
Look at the picture through the drop.
You will see lots of dots.
Newspaper pictures are made of dots.
Lift the water magnifying glass a little.
The dots look even bigger.

Amex Prices Drop
By JAMES J. NAGLE
Market Summary
AMERICAN EXCHANGE
CORPORATION BONDS
BOND ISSUES TRADED
Percentage Gains
Percentage Drops

Electric Ghost

Cut a spooky ghost out of newspaper.
Color it any way you like.
Put it flat on your wall.
Rub it hard with the side of a wood pencil.
Rub it back and forth, all over.
Take the pencil away.
Your spooky ghost sticks by itself.
Static electricity makes the paper stick.
Rubbing makes static electricity.

Reserve Statement

This Cup Hums

Cut the bottom out of a paper cup.
Cover the top with wax paper.
Hold it on with a rubber band.
Touch your lips to the wax paper.
Sing or hum into it.
Your humming makes the air move.
The air makes the wax paper move.
The cup hums along with you.

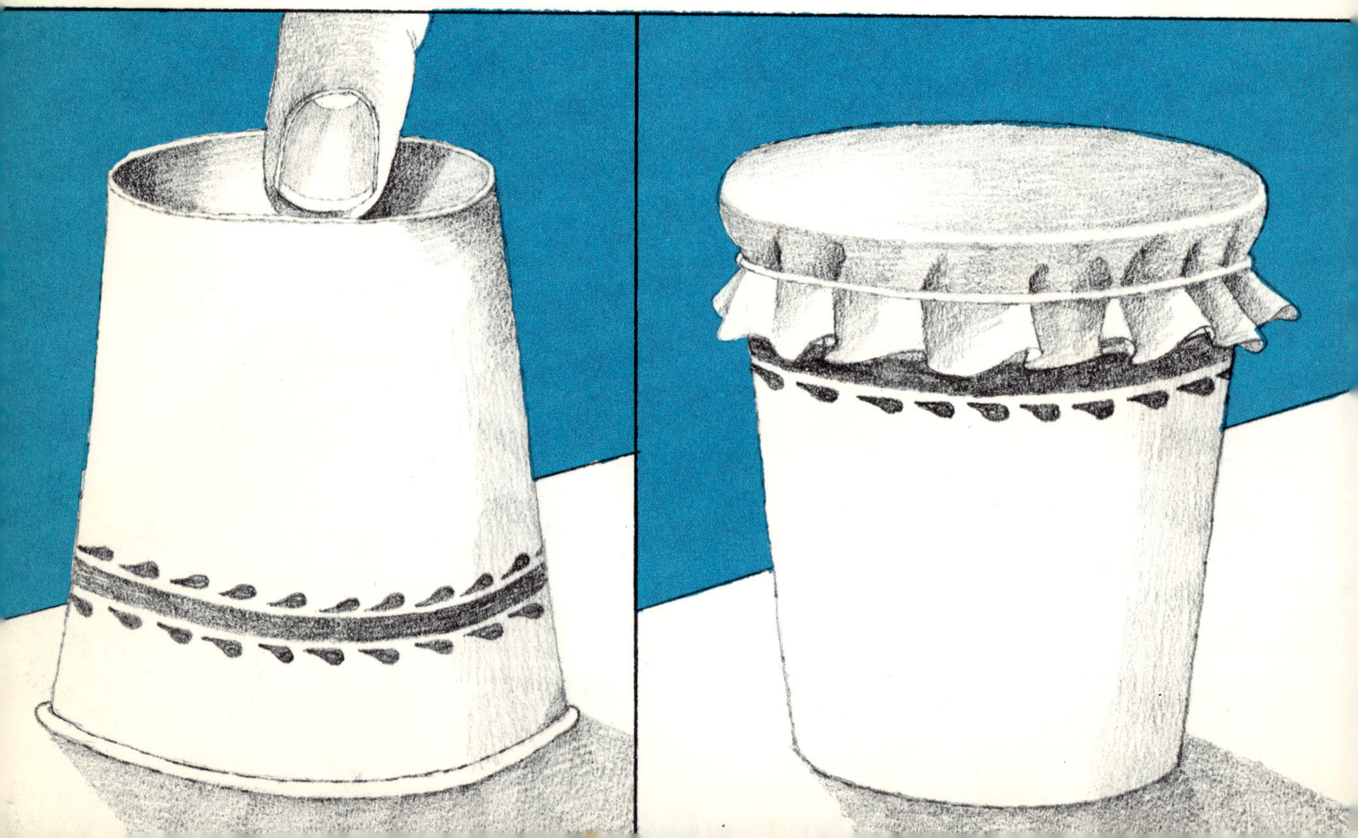

The Cup That Roars

Tie a paper clip on a piece of string.
Make a tiny hole in the bottom of a paper cup.
Poke the string through the hole.
Rub your thumbnail down the string,
while you are squeezing the string tightly.
You'll hear a "ROAR!".
The box is like a drum.
It makes the sound of your rubbing loud.
Try rubbing the string without the cup.
What happens?

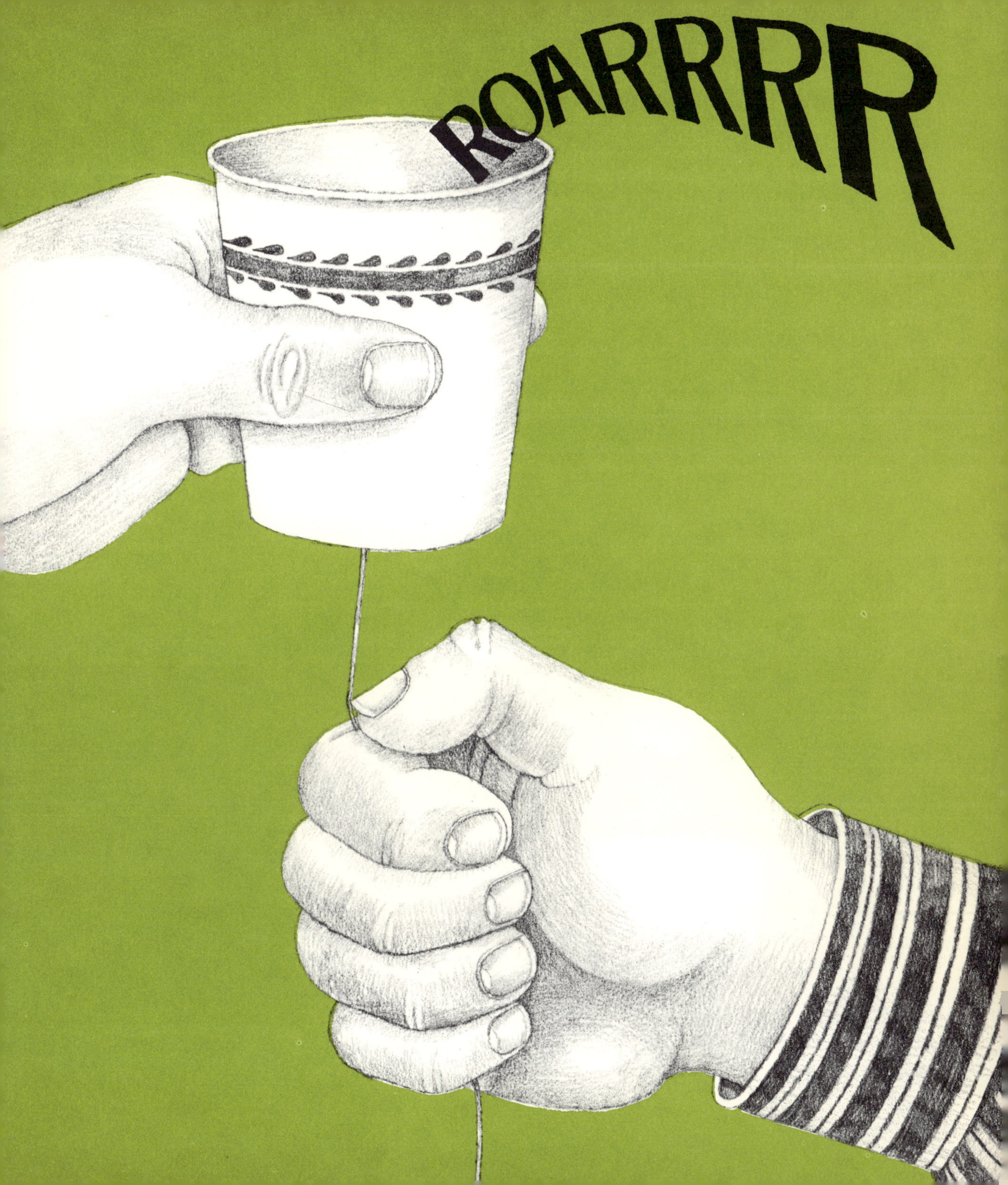
ROARRRRR

Three Note Music Maker

Put a rubber band around a book.
Slip a paper cup under the band.
Move the cup toward one end.
Snap the band.
It makes a musical note.
Snap it in another place.
The rubber band is like a guitar string.
Your music maker can make three different notes.
Keep snapping in different places.
See if you can find all three musical notes.

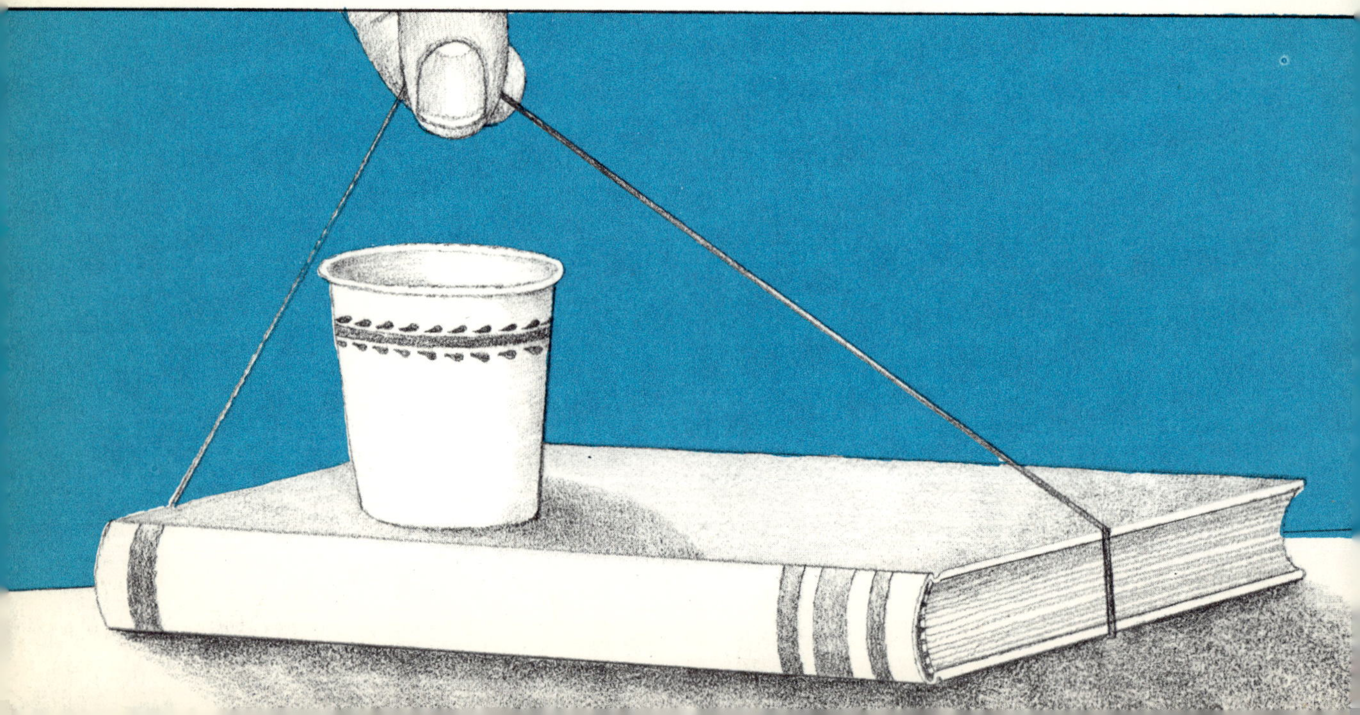

Hear a Heart Beat

Cut the bottom out of a paper cup.
Put the big end on a friend's chest.
Put it right over his heart.
Press your ear against the cup.
Do not touch the cup with your hands.
Listen . . . very quietly.
You will hear his heart beating.
Thump-thump . . . Thump-thump . . . Thump-thump.

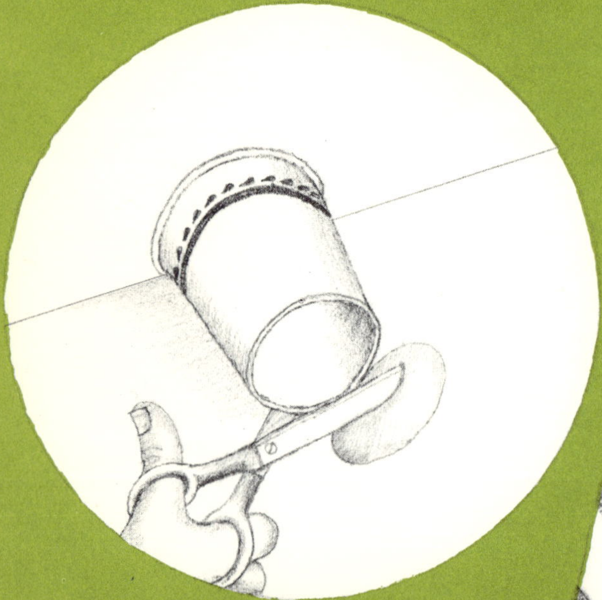

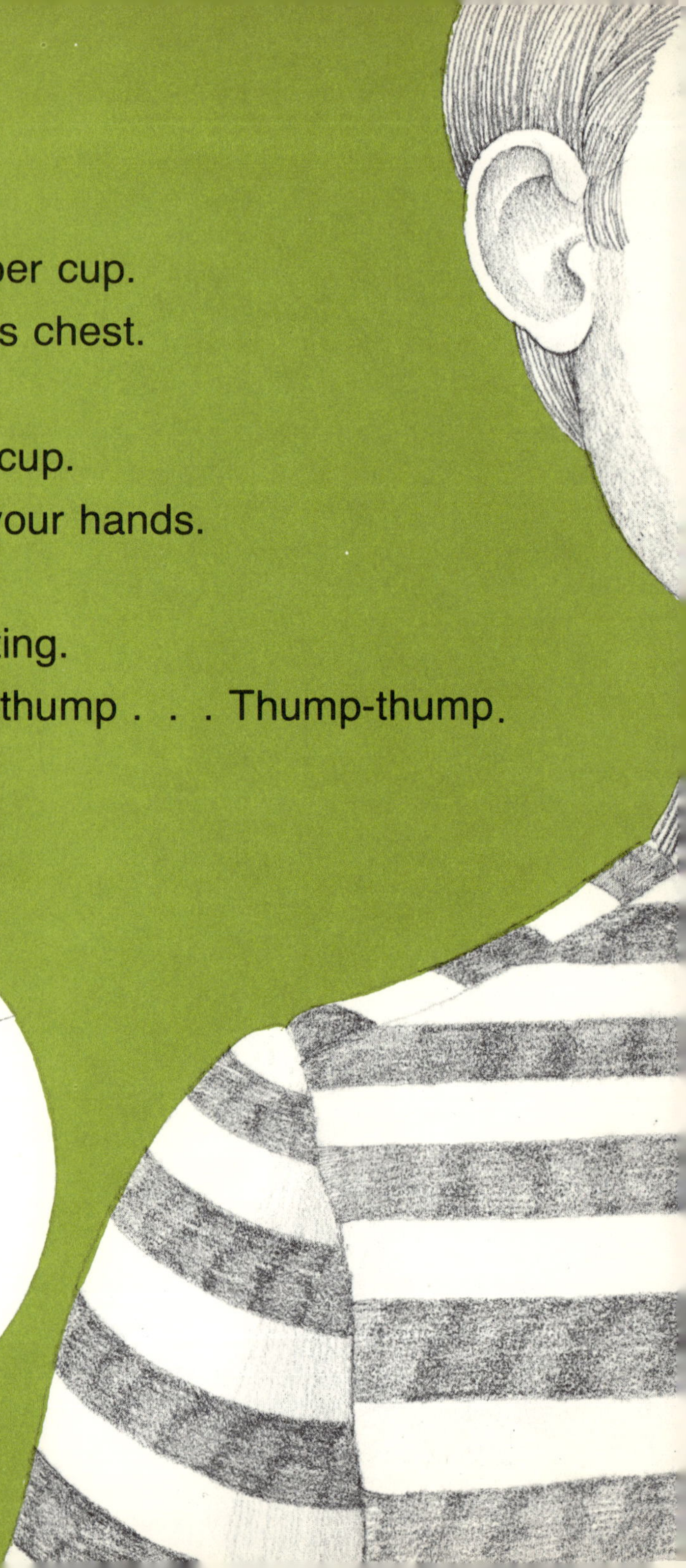

This Dog Rolls Over

Partly blow up a long balloon.

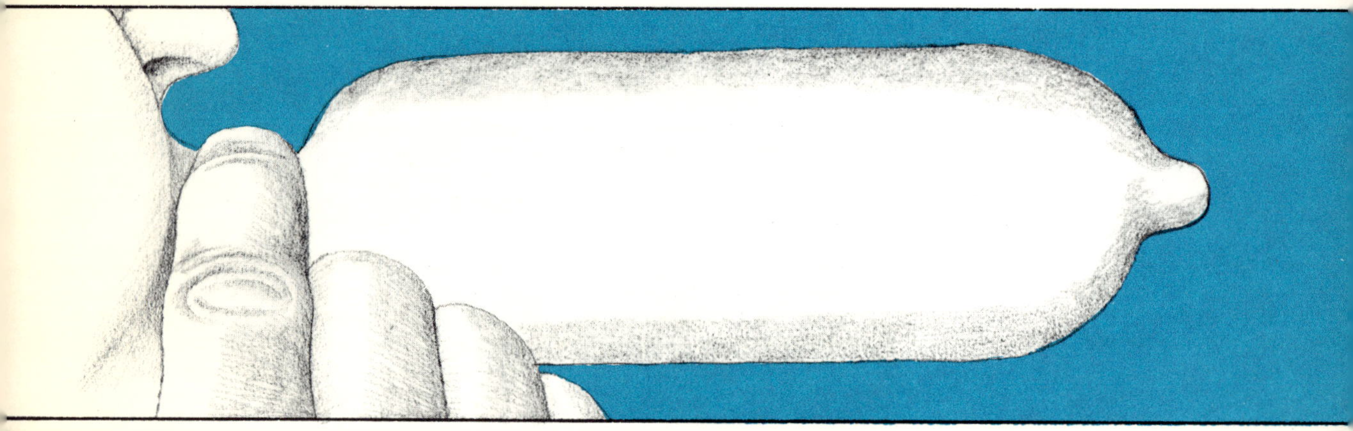

Draw a face and feet on it with a ballpoint pen.

It looks like a funny dog.

Rub his back hard in your hair.
(Be sure not to rub his stomach.)
Now set him down on the table.
Set him with his stomach down.
Let go.

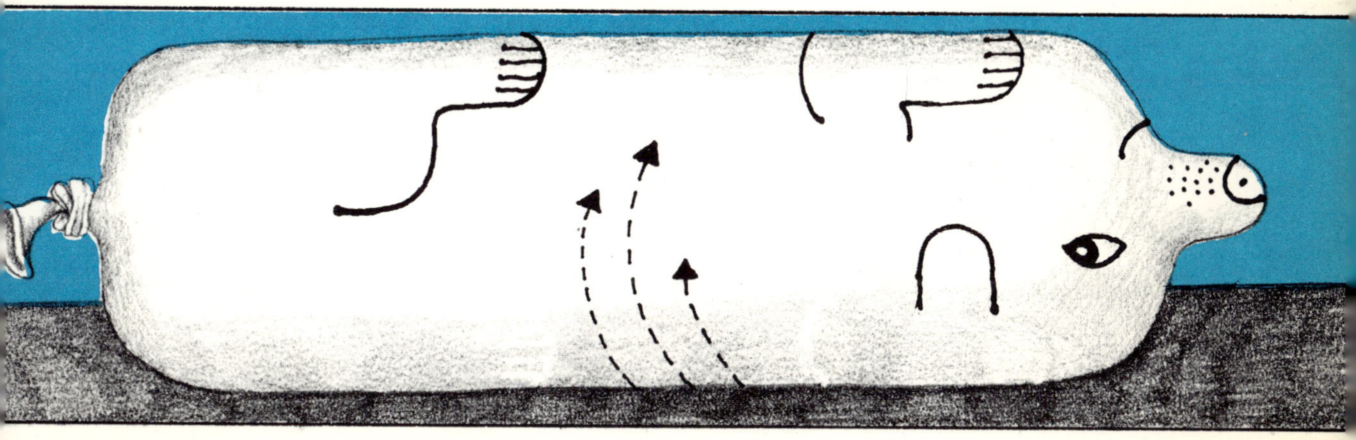

He rolls over on his back!
Static electricity from your hair,
rubbed on the dog's back,
makes his back cling to the table.

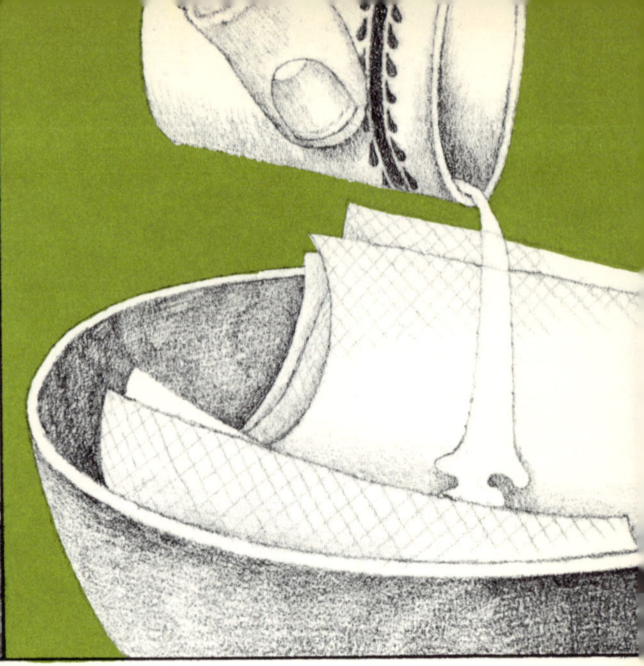

Grow a Snack Garden

Ask your mother if she has some mustard seeds.
Place three paper napkins in a dish.
Wet them with water.
Sprinkle on some mustard seeds.
Wet the napkins every day.
Do not let them dry out.
In a week tiny mustard plants will sprout.
Pick them up.
Eat them.
They are very tasty!

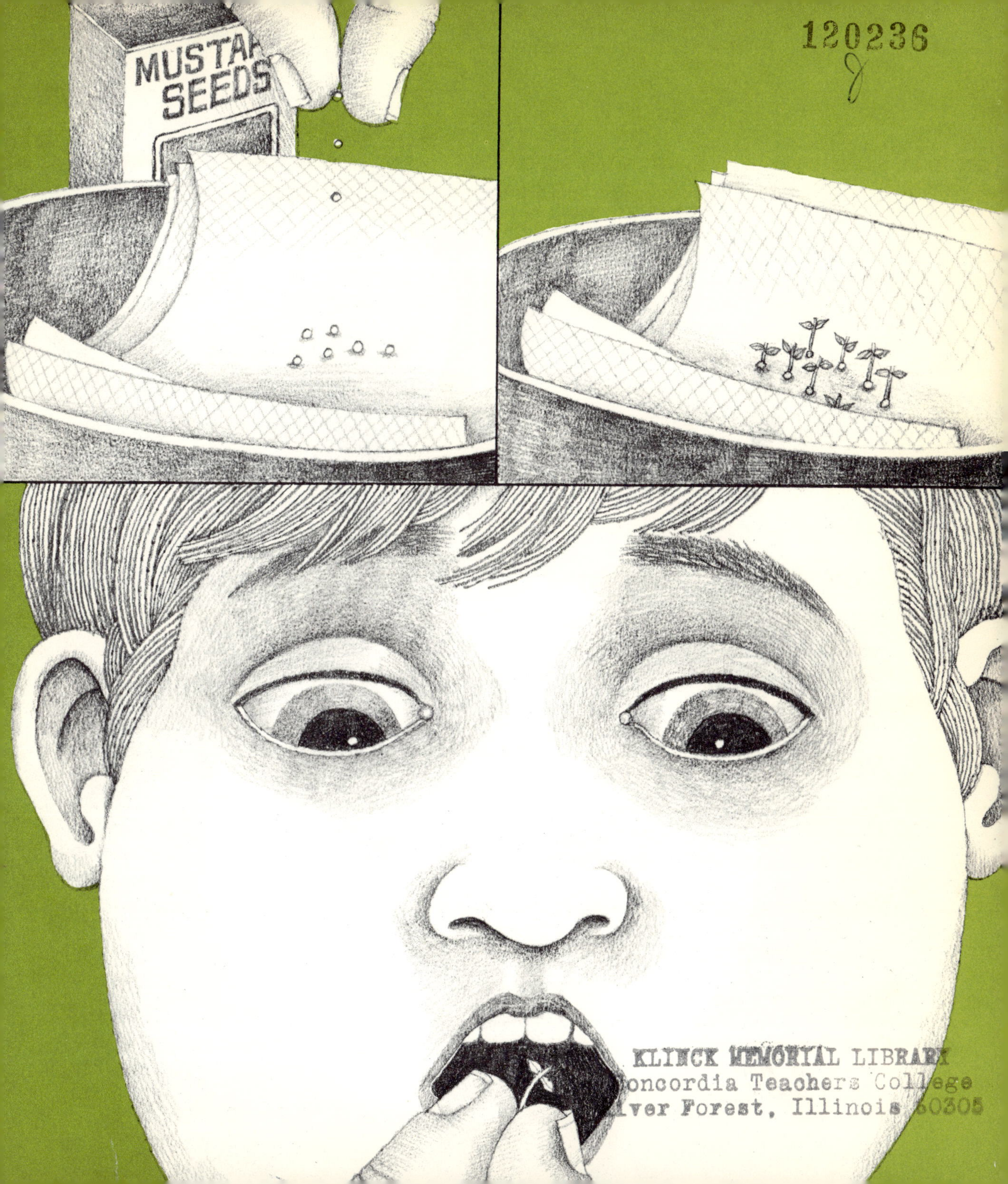
SEEDS

Draw a Glad-Sad Face

Draw this face on a paper plate.
Copy it carefully.
This plate is an optical illusion.
It can fool your eyes.
When you are happy make it glad.
When you are not happy turn it upside down.
Does it look sad?
It fools your eyes when it turns over.

Balancing Curve

Cut a paper plate in half.

Cut the middle out of one half.

Fold the curve in half.

Open it a little.

Now balance the folded curve on a pencil.

Or on your nose.

It looks like it should fall off.

It won't.

It is balanced.

Water Go Round

Push a pencil part way through a paper plate.
Push it right in the middle.
Wiggle the pencil to make the hole loose.
Take it to the sink.
Turn on the water.
Hold one edge of the plate under the water.
You now have a Water Go Round that spins!
How can you make the wheel spin faster?

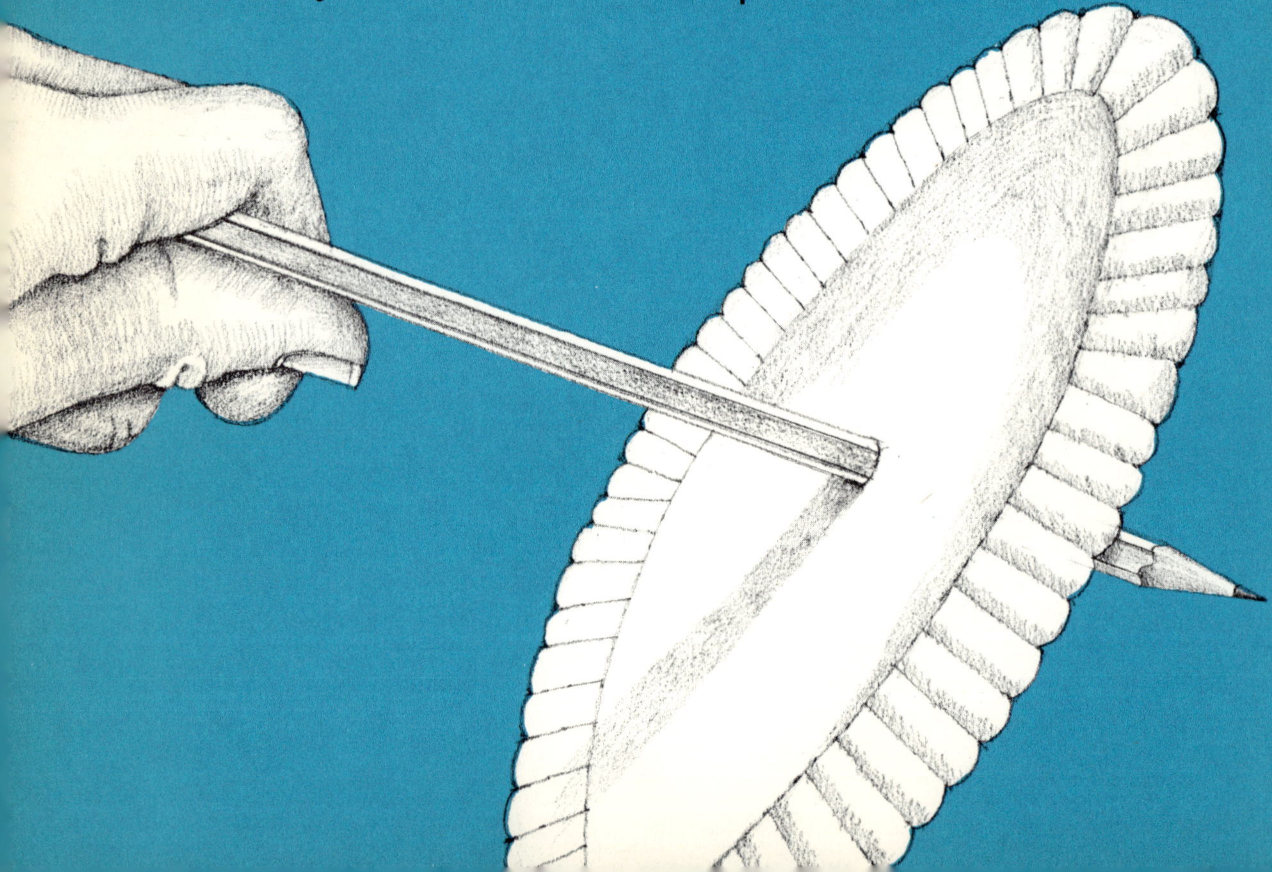

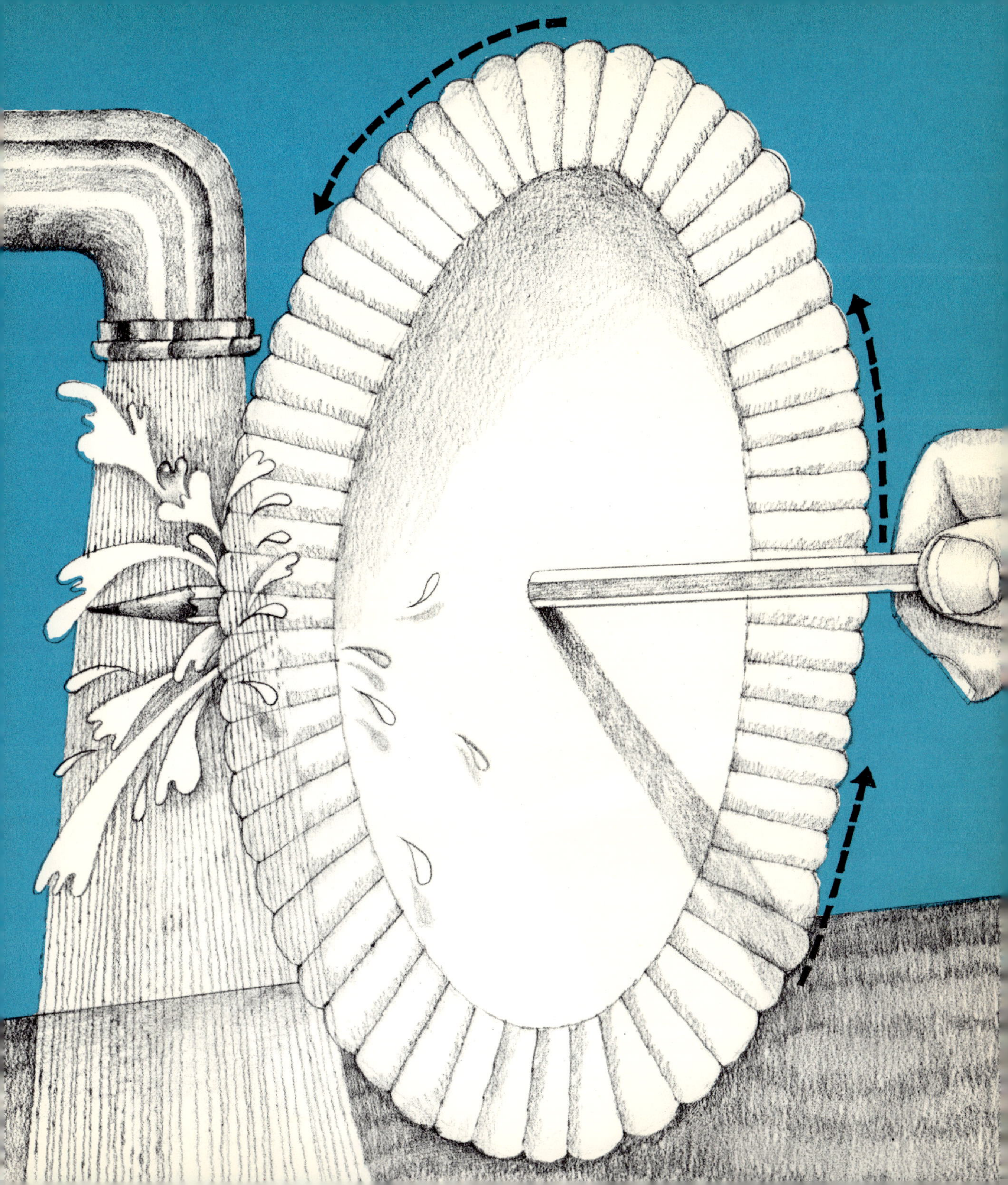

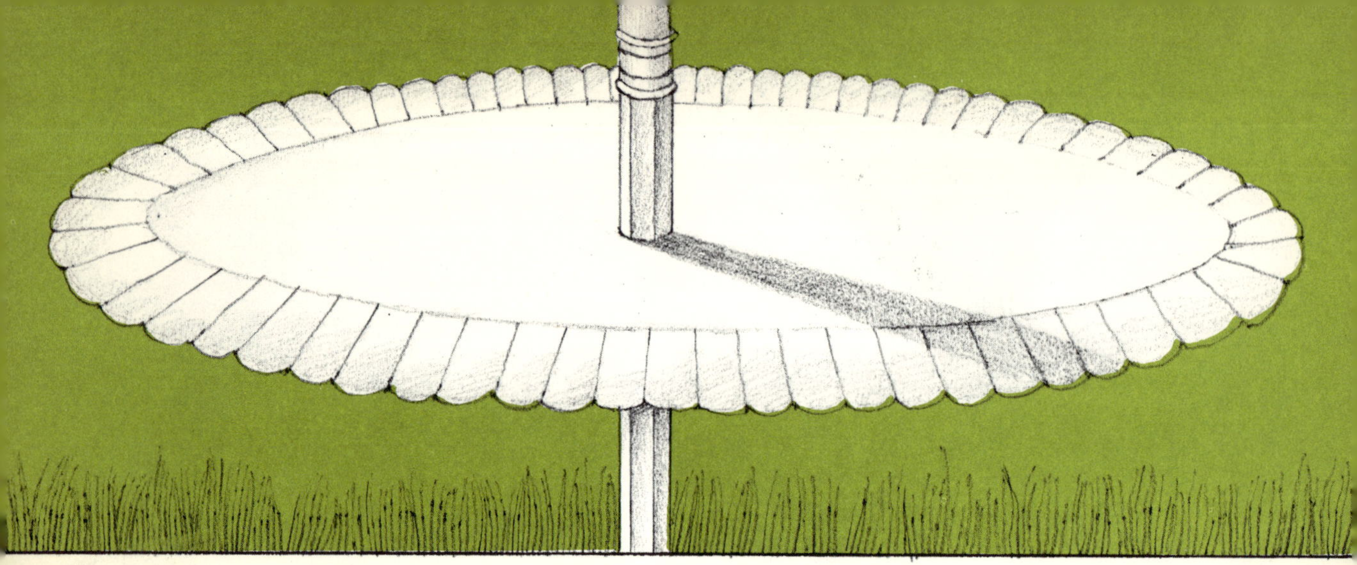

A Shadow Clock

On a bright sunny day
push a pencil half way through a paper plate.
Poke the pencil in the ground.
The pencil makes a shadow on the plate.
Make a mark where it is.

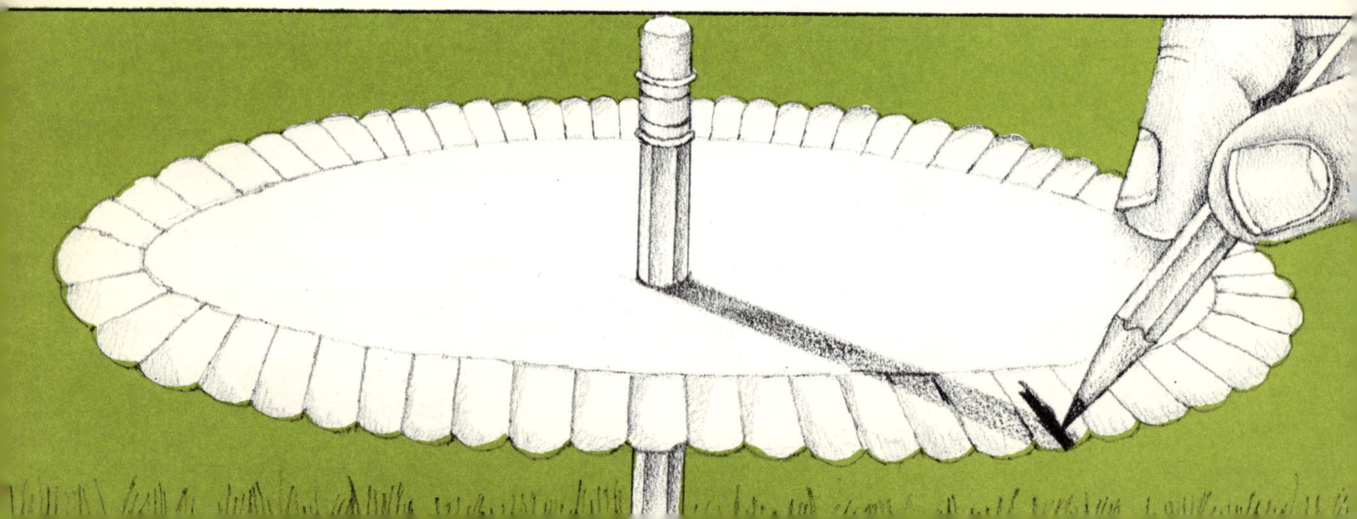

Leave it alone for an hour or more.
When you come back, where is the shadow?
The shadow has moved.

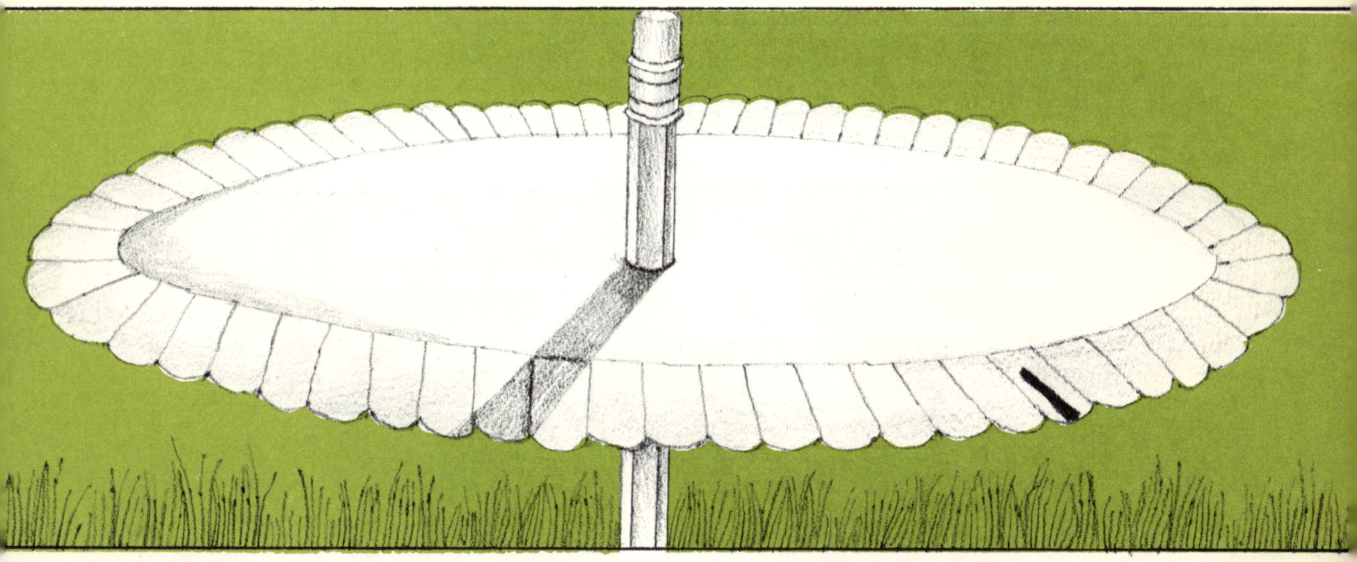

The sun has moved in the sky.
The plate is like a clock with a shadow hand.
People once used sun clocks like this.
The sun helped them measure time.

Wind Spinner

Cut a paper plate from the middle.
Fold the points back and forth.
Wait until the wind is blowing outside.
Set the wind spinner on the sidewalk.
WATCH OUT!
It will zoom away.
The wind can really move things.

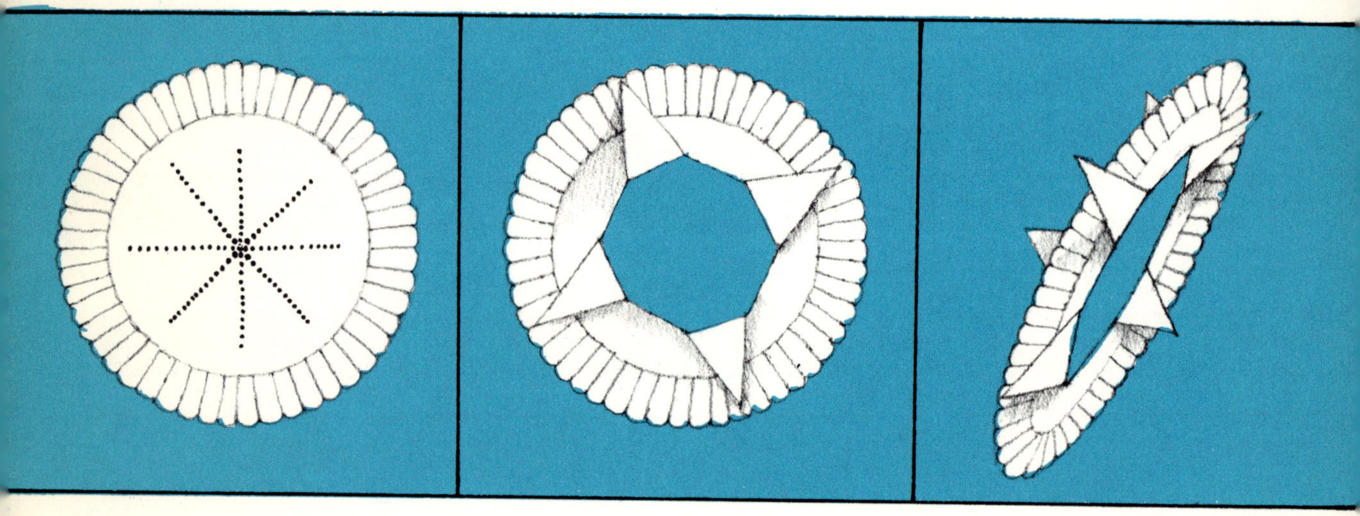

Strange Music From a String

Tie some things on a long string.
Try spoons, pencils, a ruler, and a clothes hanger.
Tie the ends of the string to two chairs.
Place your ear on one end of the string.
Push the string against your ear.
Give a friend a spoon.
Have him tap the things on the string.
Sounds travel through the string.
You hear strange music.

How to Make Raisins Dance

Pour some soda pop in a clear glass.
Drop in some raisins.
They will dance up and down.
Why? Look carefully.
Tiny bubbles stick to them.
The bubbles float the raisins up.
The bubbles disappear at the top.
Then the raisins sink.

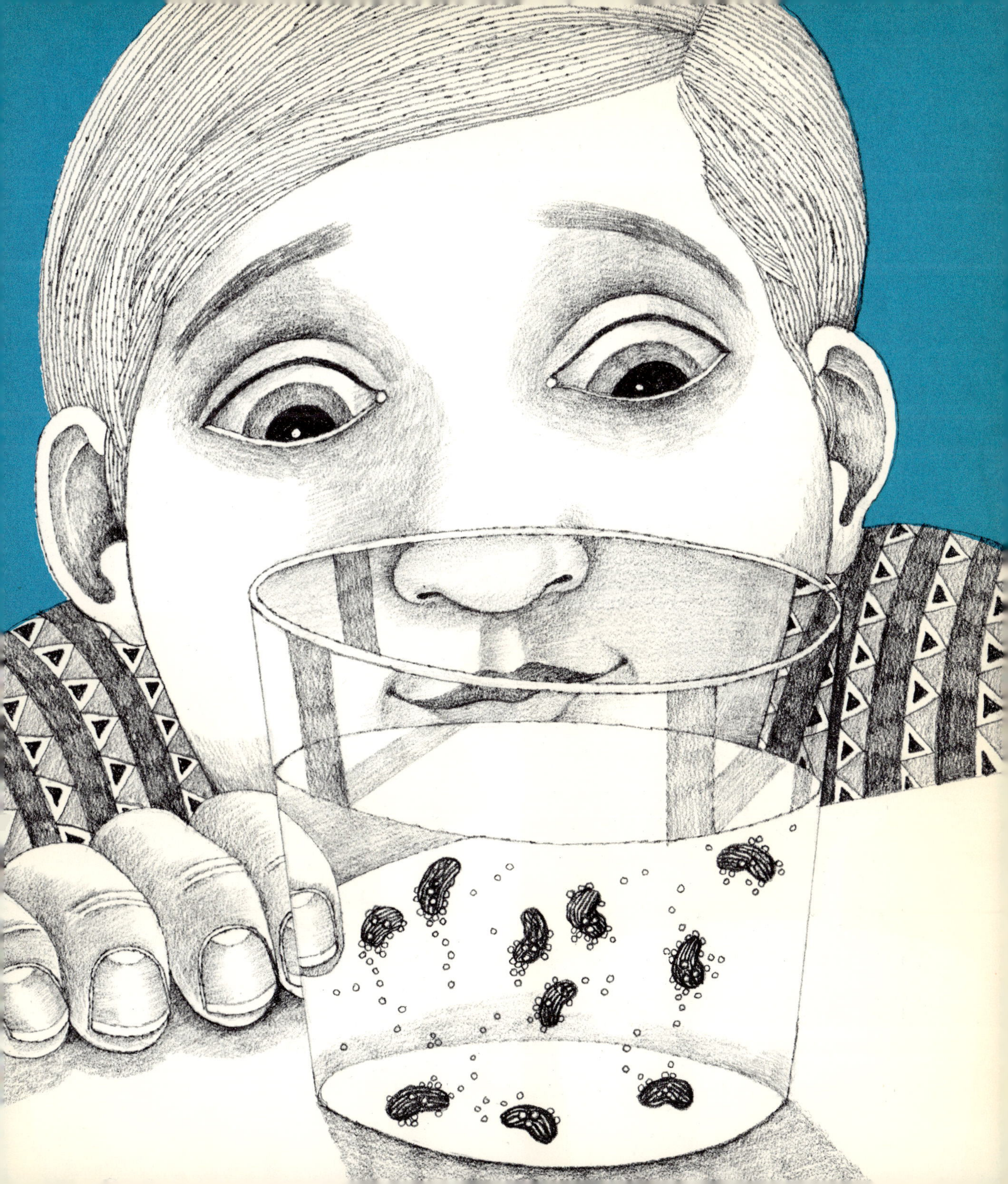

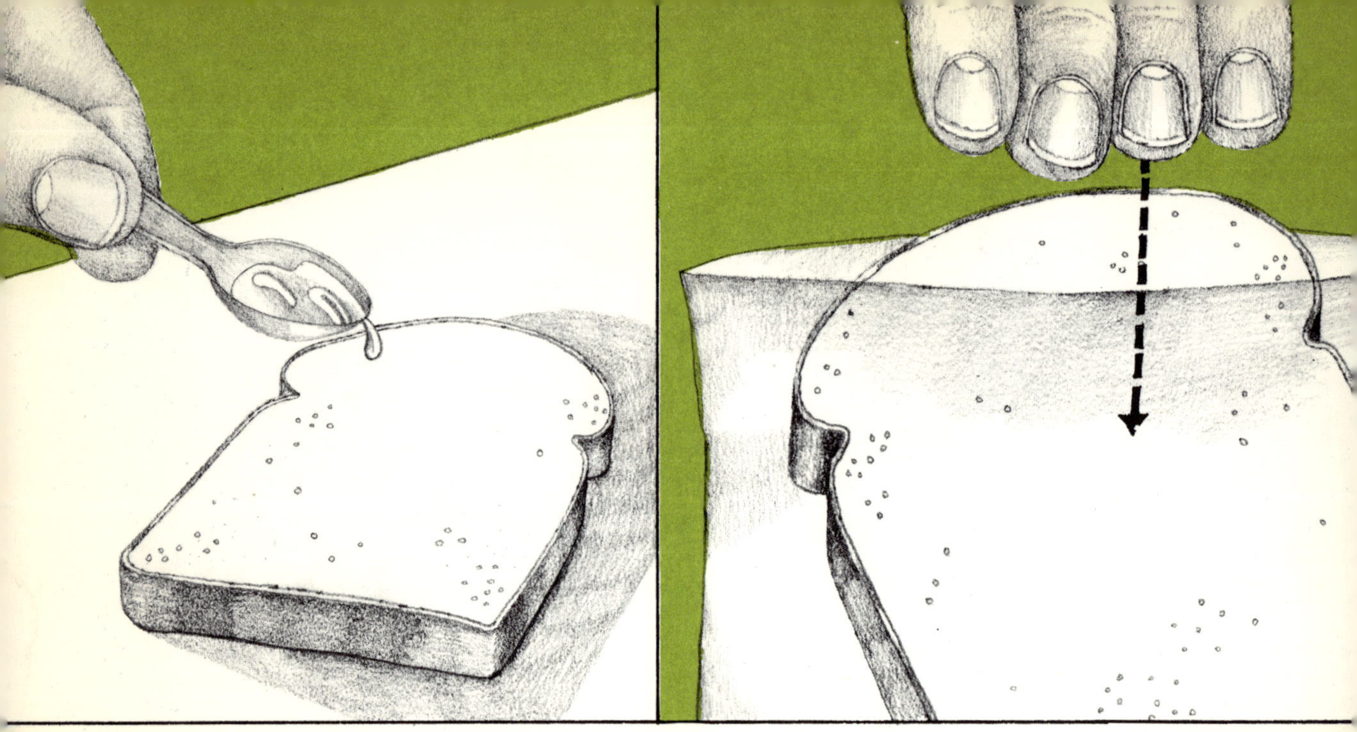

Grow a Mold Garden

Pour a little water on a slice of bread.
Put the bread in a plastic bag.
Blow up the bag.
Close the end with a rubber band.
Let it alone for a week.
Tiny mold plants will grow in the bread.
They are like miniature mushrooms.
And they are very pretty.
(Throw everything away when you are done.
Do not open the bag. It would smell!)

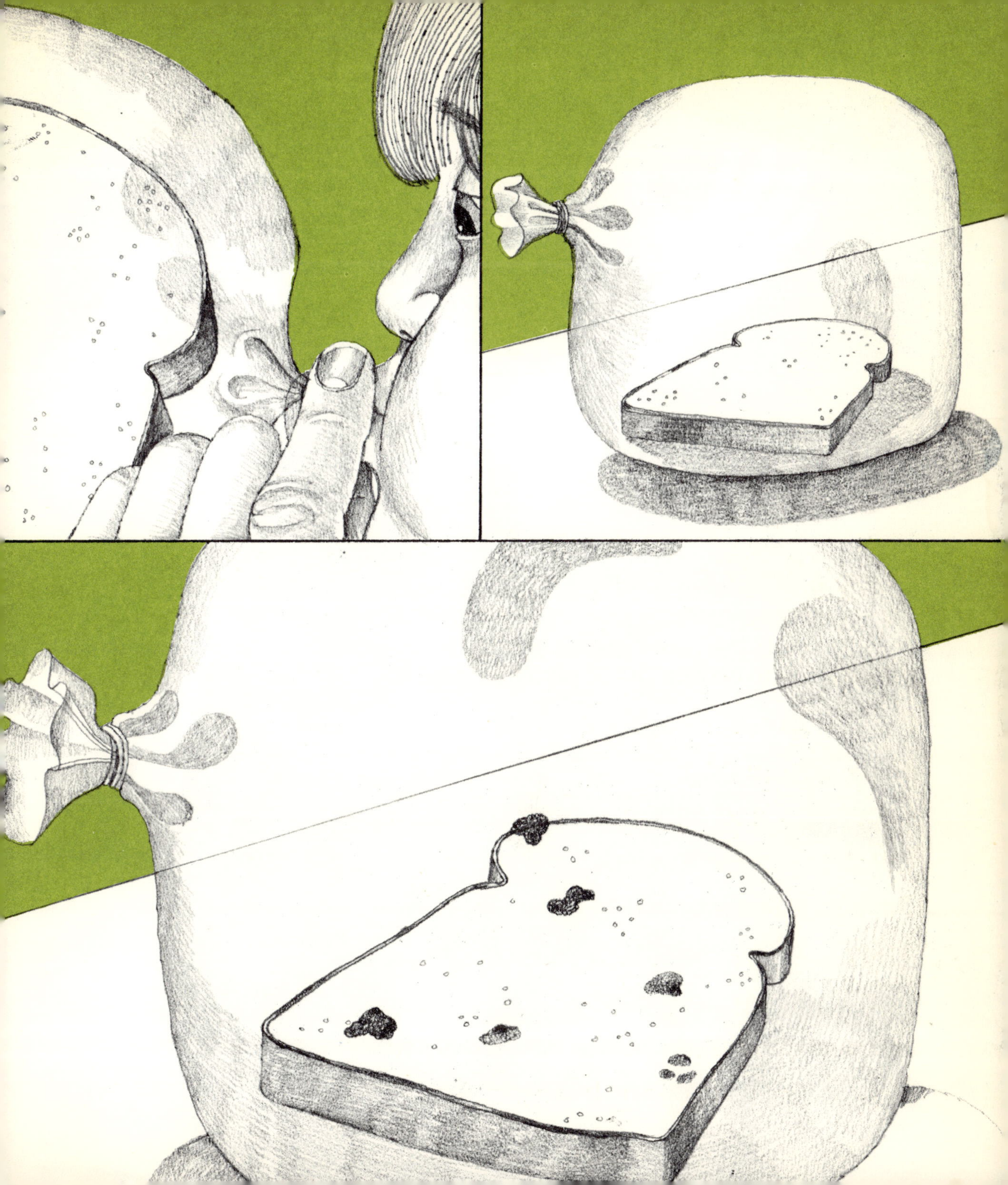

You learned about science
from the toys in this book.
What about other toys?
Can you learn any science from them?
Maybe yes, maybe no.
Play with them.
See if science makes other toys work.